Hot an
Weather

BY FRED AND JEANNE BIDDULPH

The Seasons

The photographs (below and opposite) show two seasons of the year. One shows summer. The other shows winter.

What makes the weather hot
in summer? What makes the weather cold
in winter?

We get our warmth from the sun.
The earth slowly spins around the sun.
It tilts to one side as it spins.

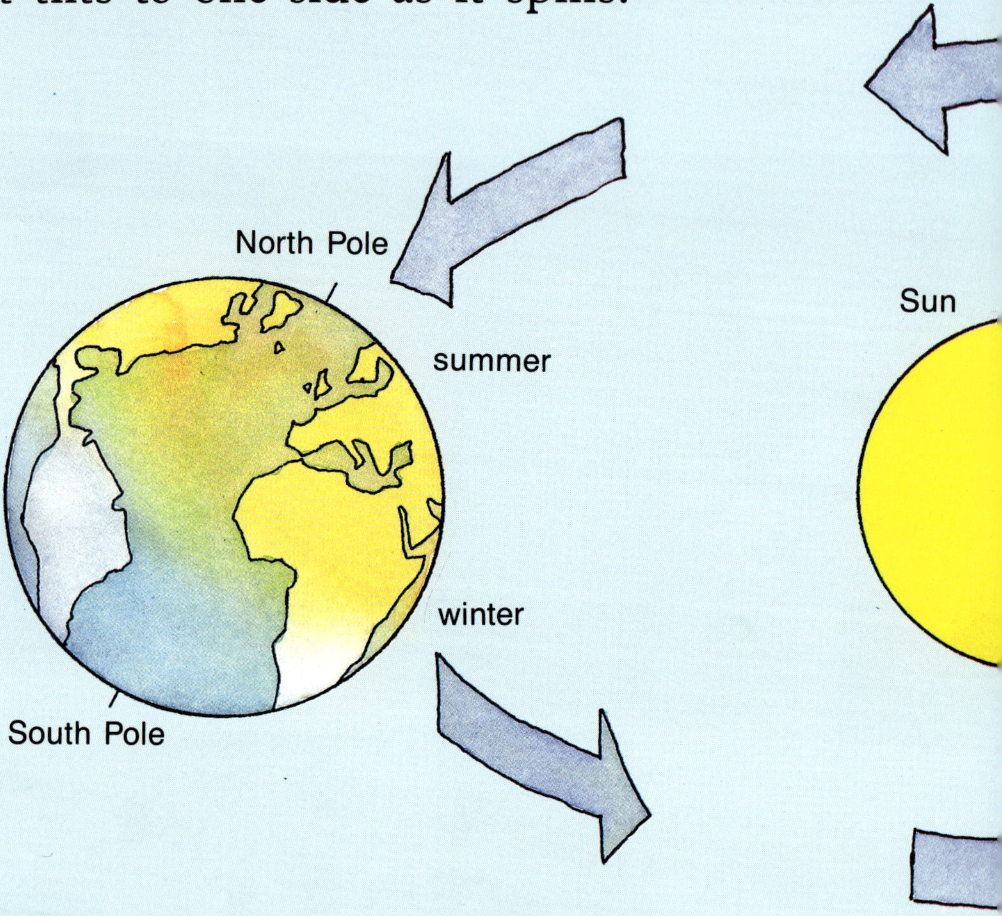

When the North Pole is tilted toward the sun, the northern part of the world has summer, and it is winter in the south.

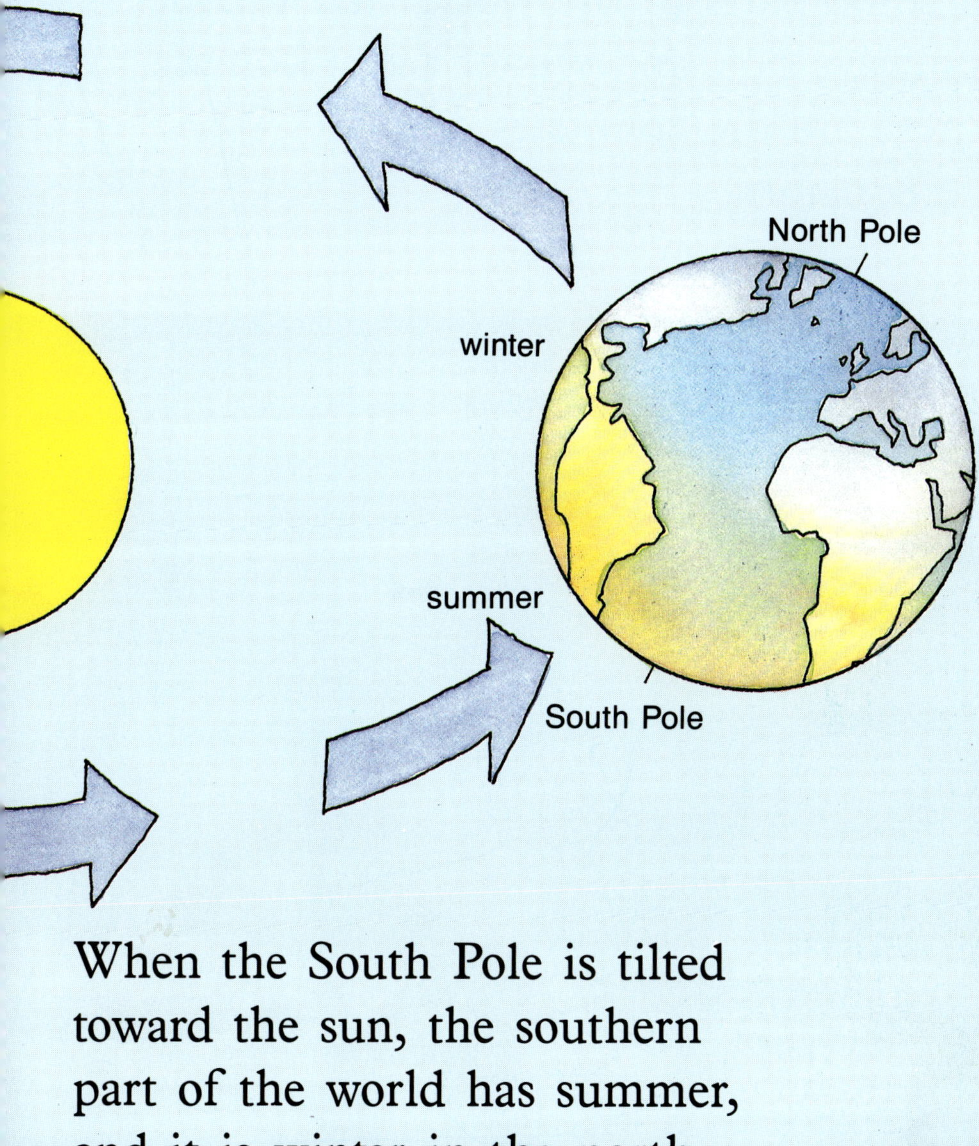

When the South Pole is tilted toward the sun, the southern part of the world has summer, and it is winter in the north.

Does the Whole World Have Summer and Winter?

In some parts of the world, the weather is always hot, and winter never comes. These are the countries around the equator.

A beach near the equator

The equator is an imaginary line that runs around the middle of the earth. It is exactly halfway between the two poles.

In some parts of the world, the weather is always cold. In these places, there is always snow and ice.

Snow and ice in Antarctica

In the Arctic, near the North Pole, and the Antarctic, near the South Pole, the sun's rays hit the earth at a lower angle than in the rest of the world.

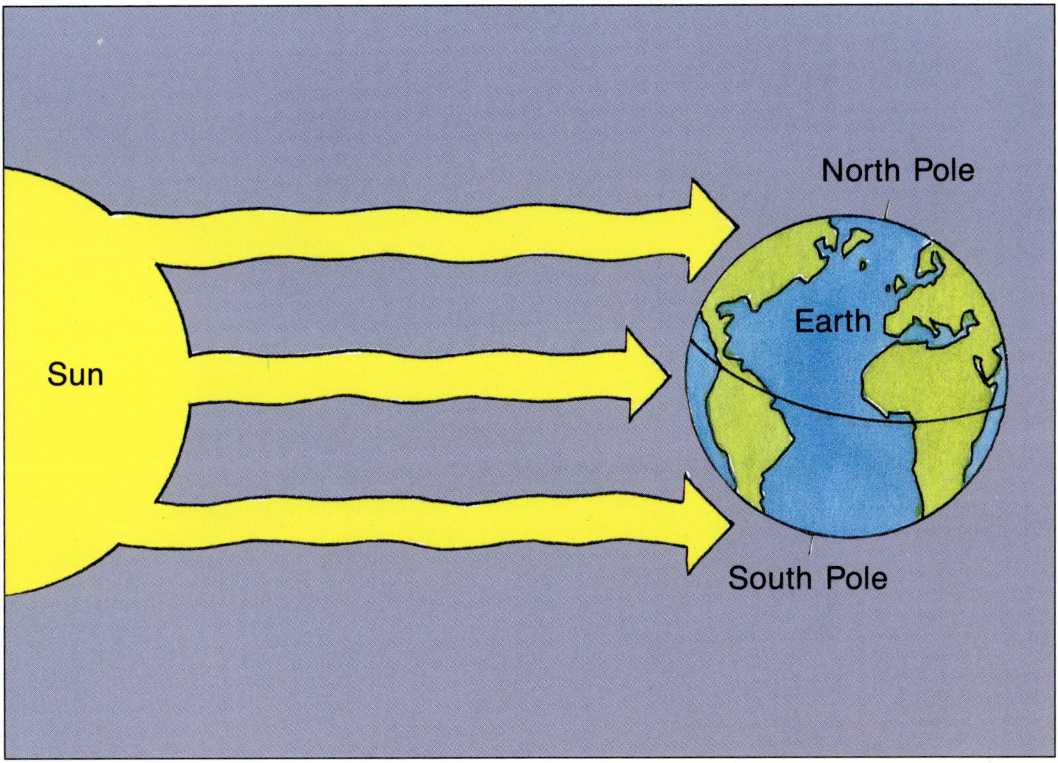

These places have short, cool summers — when the sun always shines and there is no night. They have long, cold winters — when it is always night.

Temperature and the Time of Day

Does the time of day affect the temperature?

In the middle of the day, the temperature is warmer because your part of the earth is facing the sun more directly at that time.

In the morning and again in the evening,
the temperature is cooler
than in the middle of the day.
This is because your part of the earth is
not directly facing the sun.

Temperature and the Atmosphere

The layer of air around the earth affects the temperature, too. We call this air the *atmosphere.* Without the atmosphere, it would be burning hot every day and freezing cold every night.

The atmosphere surrounds the earth.

What does the atmosphere do?

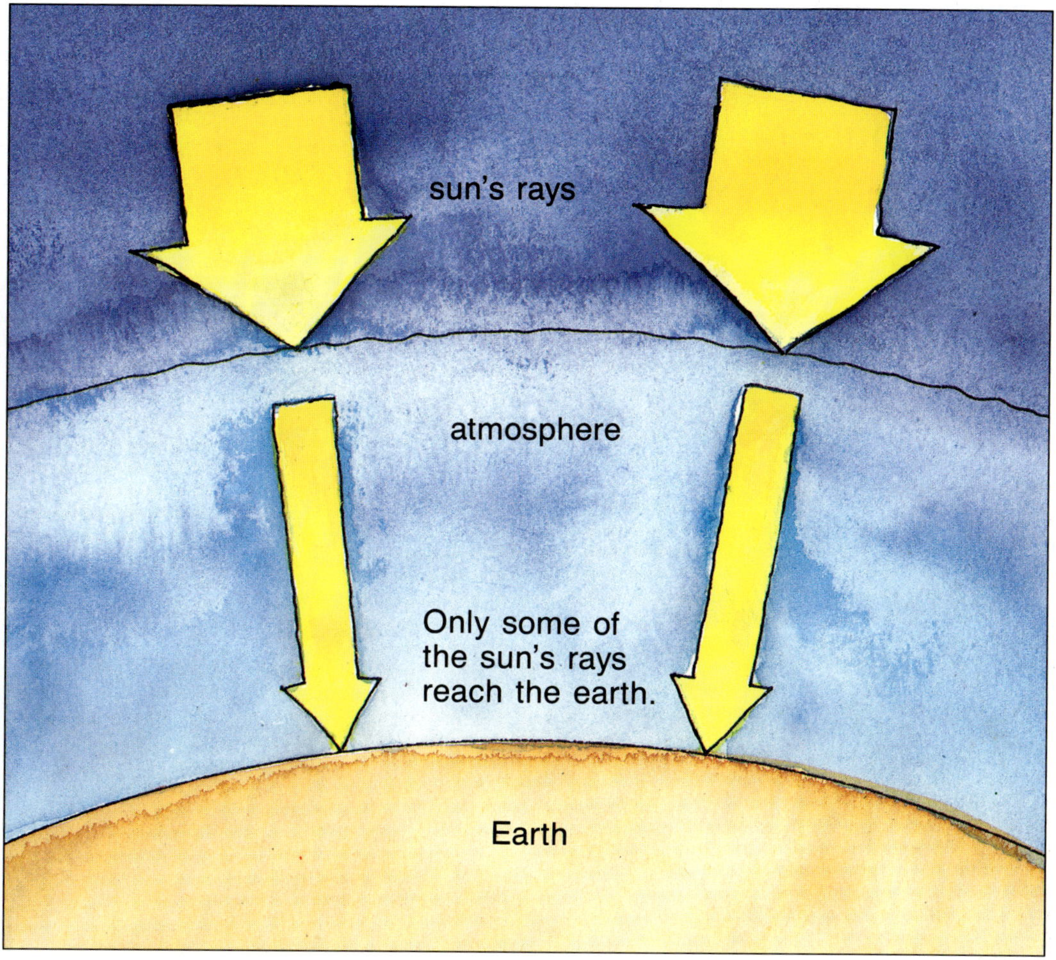

During the day, the atmosphere keeps the hottest rays of the sun from reaching earth.
Without the atmosphere, the temperature would be very hot—up to 175° F.

At night, the atmosphere keeps the warmth of the sun from escaping into space. Without the atmosphere, the temperature during the night would be very cold—about -220° F.

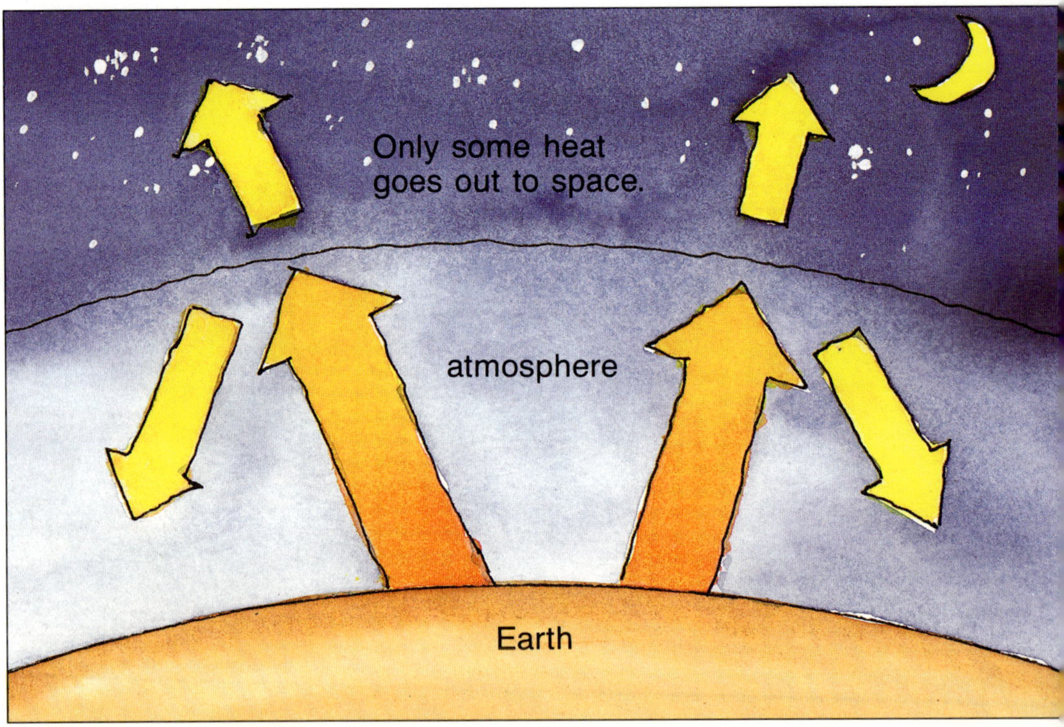

We measure temperature with a thermometer. Water freezes at 32°. This thermometer reads about 25°.

Temperature and Clouds

Clouds can affect temperature, too.

On a cloudy day, less of the sun's warmth reaches the earth. The temperature is cooler than when the sky is clear.

On a cloudy night, less heat from the ground can go up into space. The temperature is warmer than when the sky is clear.

Smoke, Dust, and Ash

Smoke, dust, and ash can make the temperature cooler.

Smoke from big fires can blot out the sun and make temperatures cooler.

Dust storms can keep the sun's heat from reaching the earth.

Ash from volcanoes can build up into clouds, bringing cooler temperatures.

What Else Affects Temperature?

Temperatures get colder the higher we go. If we climbed a mountain or took a balloon ride, we would find it much colder than at sea level.

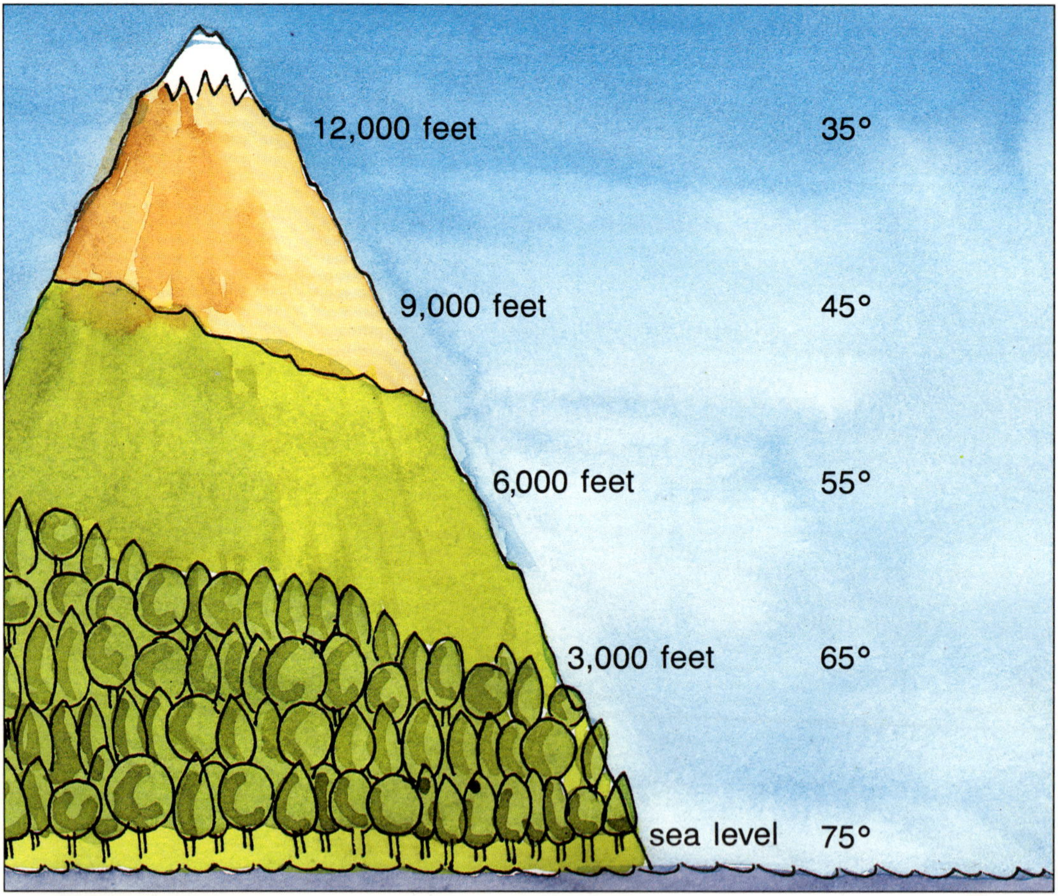

For every 3,000 feet we climb, the temperature drops about ten degrees.

Even mountains in very hot parts of the world may have snow on them all year round.

Mt. Kilimanjaro is in a very hot part of Africa, yet its peak is always covered with snow. It is 19,340 feet high.

Distance from the sea
can affect temperature, too.
The sea takes longer to warm up
or cool down than the land.
So in summer, it is warmer inland
and cooler on the coast.
In winter, it is cooler inland
and warmer on the coast.

If you lived in a hot desert,
you might find the nights very cold.
Sometimes deserts can be
freezing cold at night,
even in the summertime.
The desert sky is so clear
that a lot of heat escapes into space.

What Does Temperature Do?

We can feel it when
the weather is hot or cold.
We can see it, too.

When the air near the ground
cools down at night, we get dew.
Some of the moisture in the air
turns into little drops of water
as it reaches the ground.

In winter, the temperature of the ground may be below freezing.
When the moisture in the air reaches the ground, it freezes into frost.

In winter, rivers and lakes, and even waterfalls, may freeze.

Glossary

atmosphere the layer of air around the earth that protects us from the harsh rays of the sun

dew drops of water that appear on the ground when the moisture in the air cools at night

equator an imaginary line around the earth halfway between the North Pole and the South Pole

frost dew that freezes on the ground

temperature the way we measure how hot or cold something is